Table of Contents

Introduction 1

Background 3

The Multi-mode Authentication Framework (MAF) 4

Fingerprint Authentication – Lightweight 6

    Overview 6

    Protection 8

    Handler Implementation 9

Fingerprint Authentication – Heavyweight 11

    Overview 12

    Protection 12

    Handler Implementation 13

Management interface 15

References 17

Appendix A – Software Organization 18

Introduction

With the trend toward a highly mobile workforce, the use of handheld devices such as Personal Digital Assistants (PDAs) is growing at an ever-increasing rate. These devices are relatively inexpensive productivity tools that are quickly becoming a necessity in government and industry. Most handheld devices can be configured to send and receive electronic mail and browse the Internet using wireless communications. While such devices have their limitations, they are nonetheless useful in managing appointments and contact information, reviewing documents and spreadsheets, corresponding via electronic mail and instant messaging, delivering presentations, and accessing remote corporate data.

Manufacturers produce handheld devices using a broad range of hardware and software. Unlike desktops and notebook computers, handheld devices typically support a set of interfaces that are oriented toward user mobility. Handheld devices are characterized by their small physical size, limited storage and processing power, and battery-powered operation. Most Personal Digital Assistant (PDA) devices provide adequate memory (at least 32 megabytes of flash memory and 64 megabytes of random access memory) and processing speed (200 MHz or higher) for basic organizational use. Such devices come equipped with a Liquid Crystal Display (LCD) touch screen (one-quarter VGA or higher) and a microphone/ soundcard/ speaker, but usually lack a QWERTY hardware keypad and rely instead on a virtually displayed one. One or more wireless interfaces, such as infrared or radio (e.g., Bluetooth and WiFi) are also built-in for communication over limited distances to other devices and network access points; so too are wired interfaces (e.g., serial and USB) for synchronizing data with a more capable desktop computer. Many high-end PDA devices also support Secure Digital (SD) and Compact Flash (CF) card slots for feature expansion. Over their course of use, such handheld devices can accumulate significant amounts of sensitive corporate information (e.g., medical or law enforcement data) and be configured for access to corporate resources via wireless and wired communications.

One of the most serious security threats to any computing device is unauthorized use. User authentication is the first line of defense against this threat. Unfortunately, management oversight of user authentication is a persistent problem, particularly with handheld devices, which tend to be at the fringes of an organization's influence. Other security issues related to authentication that loom over their use include the following items:

- Because of their small size, handheld devices are easily lost, misplaced, or stolen.
- User authentication may be disabled, a common default mode, divulging the contents of the device to anyone who possesses it.
- Even if user authentication is enabled, the authentication mechanism may be weak or easily circumvented.
- Once authentication is enabled, changing the authentication information regularly is seldom done.
- Limited processing power of the device may preclude the use of computationally intensive authentication techniques or cryptographic algorithms.

Fingerprint authentication is perhaps the best-known example of a proof by property mechanism. Other classes of authentication mechanisms include proof by knowledge (e.g., passwords) and proof by possession (e.g., smart cards).

This report describes fingerprint-based authentication mechanisms involving sensor units that communicate with the device through standard interfaces supported by most handheld devices. The report provides an overview of two different types of solutions to authenticate users and provides details of the solutions' design and implementation. The first solution uses the computational capabilities of the handheld device to authenticate a user's fingerprints. The other solution uses the computational capabilities of a specialized processor to offload processing by the handheld device.

The authentication mechanisms were implemented in C and C++ on an iPAQ Personal Digital Assistant (PDA), running the Familiar distribution of the Linux operating system from handhelds.org and the Open Palmtop Integrated Environment (OPIE). OPIE is an open-source implementation of the Qtopia graphical environment of TrollTech. OPIE and Qtopia are both built with Qt/Embedded, a C++ toolkit for graphical user interface (GUI) and application development for embedded devices, which includes its own windowing system. The Familiar distribution was modified with a Multi-mode Authentication Framework (MAF) that includes a policy enforcement engine, which governs the behavior of both code modules and users [Jan03]. That framework provides the facility to add new authentication mechanism modules and have them execute in a prescribed order.

Background

Fingerprint verification is a quick and convenient method of establishing an individual's identity. Among all the biometric techniques, fingerprint-based identification is the oldest [Boe02]. A fingerprint is made of a series of three-dimensional lines, called ridges, and the spaces between them, called valleys. Features found in the unique pattern of a fingerprint's ridges and valleys are involved in the verification of an identity. Anatomic characteristics called minutiae are the locations on a fingerprint where the ridges begin, stop, fork, or intersect. Minutia extraction analyzes and identifies the key features of the fingerprint, such as the location and direction of the ridges. Some approaches use only minutiae for matching, while others include information such as the number of ridgelines between adjacent minutiae [Boe02].

When the fingerprint image is analyzed, the minutiae points are extracted and translated into a code that serves as a template. The template is initially encrypted and stored in local memory, in the scanning device itself, or on a smart card. Templates usually have a size of between 40 and 1000 bytes, often around 256 bytes [Boe02]. All of the details of the original fingerprint cannot be recreated from the minutia data stored on the template. However, an artificial fingerprint exhibiting that minutia can be generated [Ulu04].

Authentication by means of fingerprint recognition is based on matching the features of a live fingerprint against those of enrolled fingerprints held in a data store. The technique relies on a sensor to capture an image and the necessary algorithms to perform feature extraction and matching. During verification of a fingerprint, an image of a live fingerprint is captured and translated into a template of minutiae, and then compared with the stored templates of images enrolled by the user. Authentication is successful and an identity established when the two match. Several technologies exist that can be used to obtain a digital image of the fingerprint, including capacitance, thermal, and optical sensing [Boe02].

Fingerprint authentication on a PDA is challenging since the algorithms cannot impose too high a computational demand to be impractical, yet the result must be effective with reasonable false acceptance and rejection rates. Fingerprint readers are beginning to appear as built-in hardware on some PDA devices. For example, as early as 2002, IBM & CDL (Consumer Direct Link) announced the Paron MPC PDA containing a small touch sensor, and at about the same time, the HP's iPAQ H54xx series was released containing a small swipe sensor beneath the navigation button. Fingerprints readers have also been integrated into sleeves for iPAQ and other devices.

The Multi-mode Authentication Framework (MAF)

MAF was developed previously in a related effort to provide a structured environment for the protection and execution of one or more authentication mechanisms operating on Linux handheld devices [Jan03]. The authentication mechanisms described in this report were implemented specifically for this framework. Each authentication mechanism consists of two distinct parts: an authentication handler and a user interface (UI) for the handler. Figure 1 illustrates these elements within a Linux operating system environment, enhanced with kernel support for MAF.

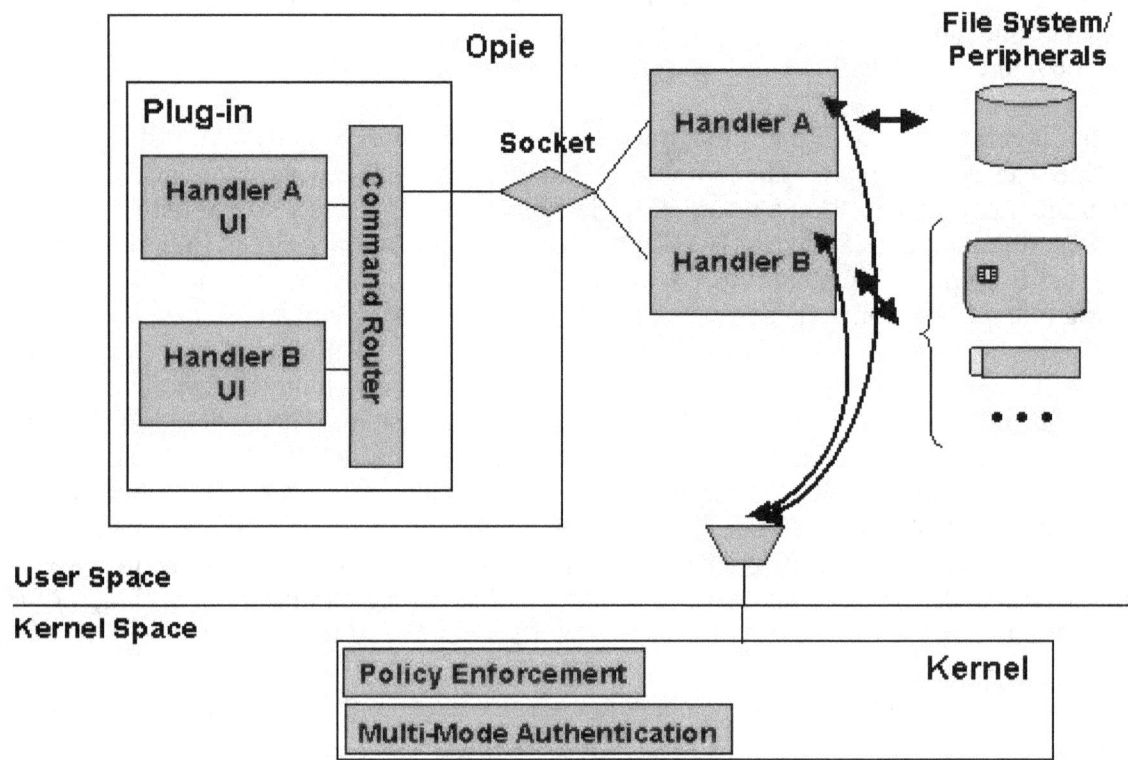

Figure 1: Multi-mode Authentication Framework

Authentication handlers embody the mechanism that performs the actual authentication. They communicate with the kernel, listening for when to initiate authentication and reporting whether authentication was successful. They communicate with the user interface components to bring up specific screens, accept input, display messages, etc. on the device. Handlers also communicate with any peripheral hardware devices needed for authentication, such as a security token, and access the file system to store and retrieve information as needed. Handlers run in user space as do their respective user interface.

The user interface for an authentication mechanism is implemented as a set of components of a plug-in module for the OPIE desktop environment. Their function is to perform all necessary interactions with the user. For example, with smart card applications they can be used to prompt for and accept entry of a personal identification number (PIN), or to notify the user of errors. The plug-in module supports a socket interface to receive commands from an authentication handler that runs as a separate process, and to route the commands to the correct user interface component. Similarly, reverse routing is also supported for responses from user interface components to an authentication handler.

4

The kernel has two key modifications to support the framework: the policy enforcement functionality and the multi-mode authentication functionality.

- Policy enforcement's main responsibility is to impose different sets of policy rules on the device, as signaled by multi-mode authentication, for one or more defined policy contexts referred to as policy levels. For example, it can block hardware buttons and certain I/O ports on the device until the user is authenticated at the lowest policy level, policy level 1. Policy enforcement is also used to protect authentication information files, the user interface and handler components, and policy enforcement information against improper access. Moreover, it also has the means to register and start up authorized handlers, if they are not running, or restart them, if they terminate for some reason.

- The main responsibility of the multi-mode authentication functionality within the kernel is to govern the authentication steps as they relate to the various policy levels that are configured. Communication between the kernel and an authentication handler is done via the /proc file system. The multi-mode authentication functionality maintains complete knowledge about the mappings between authentication mechanisms and policy levels, simplifying the development of the authentication handlers. One of its key functions is to initiate user authentication when the device is powered on. It also controls the order and frequency in which the handlers are awakened from suspended state and begin execution, and ensures that messages from only legitimate handlers are accepted and processed.

Together, the kernel policy enforcement and multi-mode authentication extensions are essential for securing authentication applications.

To create an authentication mechanism, a developer needs to create an authentication handler for the mechanism along with any needed user interface objects and the associated policy rules to protect the mechanism. Policy rules include limiting access to any storage objects used, the user interface objects within the plug-in module, and the handler itself; and limiting communications to peripheral devices and among the handler, the user interface, and kernel. Note that writing an authentication mechanism that neither interacts with the user nor requires a user interface component is possible. For example, the mechanism could be based on a sensor that is continually monitored and whose input triggers both an authenticated or non-authenticated transition.

Fingerprint Authentication – Lightweight

The lightweight fingerprint authentication mechanism relies on a fingerprint identification unit to perform the entire authentication process and return a score for verdict determination to the PDA. The work left to the PDA is mainly governing the fingerprint unit's capture and comparison of enrolled fingerprints with live prints. In addition, the PDA work includes the display of user interface messages to guide the enrollment and management of fingerprints, the determination of a pass/fail verdict from the matching score received, and the notification of an authentication success or failure to the user. Figure 2 illustrates the functional organization of the process, split between the fingerprint unit and the processor unit of the PDA. The term lightweight is used to describe this approach, since much of the work is offloaded from the PDA to the fingerprint unit, lightening the load on PDA processor unit.

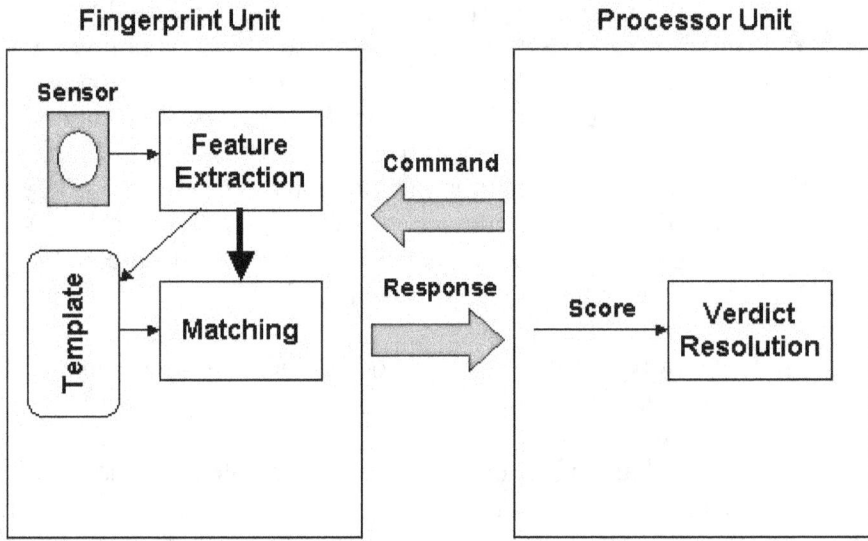

Figure 2: Lightweight Functional Organization

The Fingerprint ID Unit (FIU) 300 from Sony was used for the prototype implementation.[1] It is a two-module sensor board and verification board, with a 128 x 192 pixel, solid state capacitive sensor, 16-bit microcomputer, 1 megabyte of read only flash memory, 32 kilobytes of random access memory, and supports RS-232 and Universal Serial Bus (USB) communications interfaces. Extracted templates use 512 bytes per print, allowing up to 1000 fingerprint templates to be stored in on-board memory. Templates can also be exported and stored elsewhere. A temporary session key is used with a nonce to encrypt communications to/from the unit and protect against replay. Encryption uses the Data Encryption Standard (DES) 56-bit data Electronic Code Book (ECB) mode. Session keys are managed by the PDA, using a shared master key.

Overview

The fingerprint handler for the lightweight solution, as all MAF handlers, runs in user space. The handler manages the mechanism by communicating with the kernel, the fingerprint reader, and the Opie plug-in containing its user interface components. It guides the placement and

---

[1] More information about the FIU-300 can be found at
http://bssc.sel.sony.com/Professional/puppy/files/SONY47653_FIU300.pdf

removal of fingers during fingerprint scans and controls all the necessary steps regarding the authentication process and verdict determination.

The handler is triggered by the MAF enabled kernel when the user tries to access a specific level that the handler protects, and after obtaining a live fingerprint scan of the user, replies to the kernel with the result of the authentication (success or failure). The fingerprint handler is a non-polling one – once a user is authenticated, the mechanism does not need continual fingerprint authentication checks, as long as the policy level remains at that of the mechanism or higher. The handler uses the Opie plug-in to tell its UI to display informative messages to guide the user through the sign-on and enrollment steps. The type of interaction with the fingerprint reader depends on the device interface used (e.g., serial, USB, etc.) and the command set of the reader. In the case of the FIU-300, it uses the on-board capabilities of the device to acquire the fingerprint images and templates, control the matching process, and obtain the scores of matches.

Fingerprint authentication has two main parts: enrollment and verification. Figure 3 gives an overview of the fingerprint authentication process. The upper half shows the verification functionality and the lower half the enrollment functionality. Note that discretion to enable the authentication mechanism is, by default, left to the user. However an organizational administrator can pre-enable the mechanism by enrolling the user at the time the device is issued.

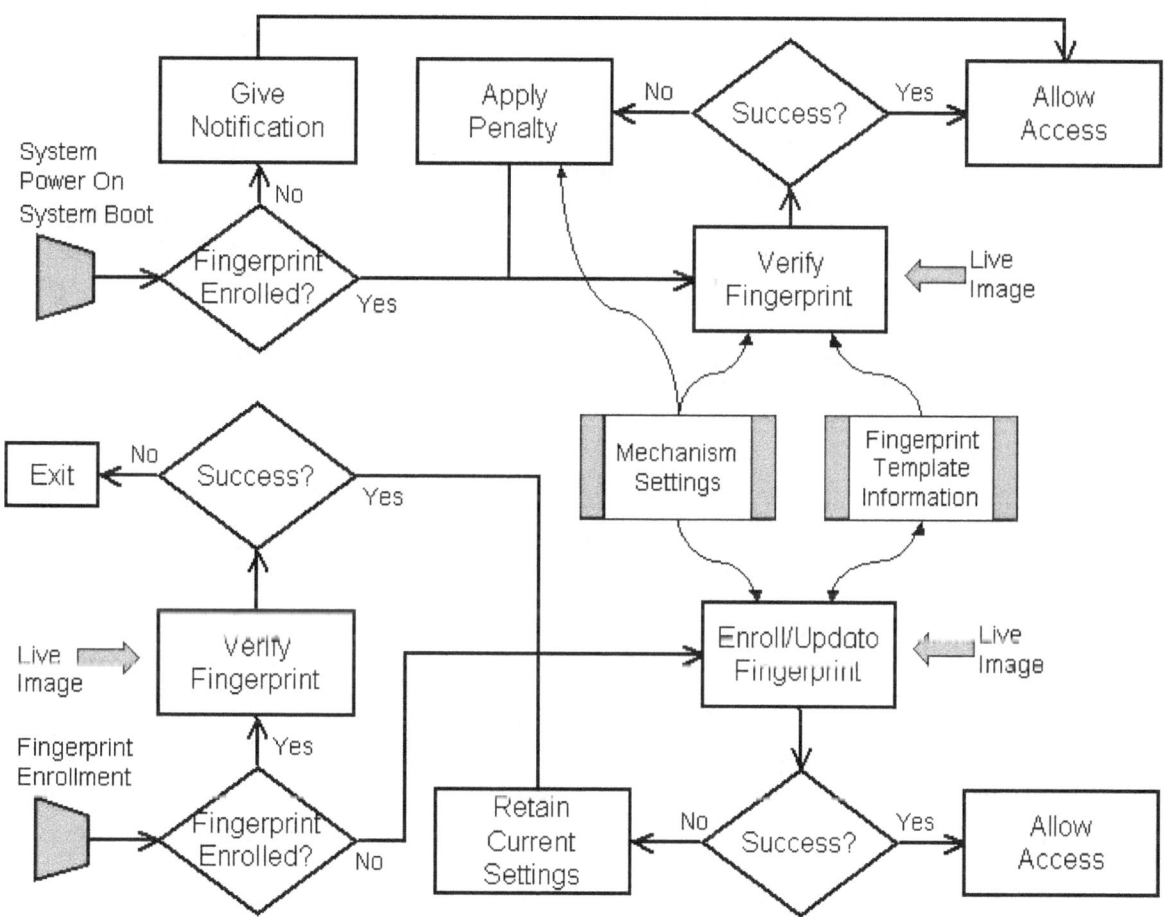

Figure 3: Authentication Process Overview

7

When activated, the handler mechanism prompts the user to login or, if fingerprints are not yet enrolled, notifies the user to do so and then allows access. The authentication mechanism is activated at both device power on and system boot up. Enrolling a fingerprint requires the device to register a fingerprint scan of the finger several times to ensure accuracy. After the user enrolls one or more fingerprints, he must successfully authenticate himself in the future before being allowed to access the device or to add new or replace existing fingerprints. If a problem arises during enrollment, the user can continue the process until successful.

Enrollment uses several types files. The mechanism settings file contains information related to the scanning process, such as the number of fingerprints to collect and the root shared key for encrypting information to and from the fingerprint unit. When a successful enrollment occurs, the resulting template derived from the fingerprint scans taken of the user are saved away within the fingerprint template information file, and the user gains access to the device. Note that for the lightweight variant, fingerprint templates are maintained in the memory of the fingerprint unit.

Once enrolled, subsequently powering on or booting up the device prompts the user to provide a live fingerprint image for verification. The verification process uses information from the mechanism settings file to encrypt information between the device and the fingerprint reader. A correct match against the enrolled fingerprint templates in the fingerprint information file results in successful authentication and access is granted to the device. If too many authentication failures occur, further attempts are blocked temporarily to prevent unrestricted password guessing attempts.

Any time after gaining access, a user can update the set of enrolled fingerprints by using an available icon to launch the process and providing a live fingerprint image for verification, which then follows the same procedure described above for verification at power on or boot up. In Figure 3, the "Verify Fingerprint" boxes associated with fingerprint update do not show the information flows discussed above, but are present implicitly. Successful verification allows the user to enroll new fingerprint images adding to or replacing the existing set. A successful enrollment updates the fingerprint template information file and the user regains access to the device.

Protection
For user authentication the fundamental threat is an attacker impersonating a user and gaining control of the device and its contents. Fingerprint units should be embedded into devices that are designed to resist physical tampering and avoid exposing the communications channel between the device and the unit. Presuming those safeguards are effective, the following vulnerabilities are the main candidates for exploitation:

- The authentication mechanism can be bypassed
- Weak authentication algorithms and methods are used
- The implementation of a correct and effective authentication mechanism design is flawed
- The confidentiality and integrity of stored authentication information is not preserved

The fingerprint handler uses the encryption capabilities of the FIU-300 to protect its communications with the unit. The openssl library is used to carry out the encryption operations on the PDA. Fingerprint templates are maintained on the FIU-300. Storing the templates in the memory of the biometric device avoids risks associated with their transmission [Pol97].

The lightweight fingerprint authentication mechanism relies on MAF, which in turns relies on the security of the underlying operating system implementation. The handler must be protected from substitution and overwrite respectively through the multi-mode authentication and policy enforcement functionalities of MAF. Substitution is prevented through an entry in the list of registered handlers (e.g., </usr/bin/handlerSMMC 2>) identifying its location, while overwrite is prevented with the following policy rules in the MAF policy file (/etc/MAF/defaultPolicy), which also grants exclusive access to information maintained by the handler:

<file   /etc/MAF/FP*   /usr/bin/HandlerFP   0>
<file   /root/Settings/FP/*   /usr/bin/HandlerFP   0>

Handler Implementation

The lightweight fingerprint handler operates as a non-polling handler, allowing a prescribed number of fingerprint scans before giving up. The following code excerpt shows the main execution loop of the handler, during a regular authentication (device available, fingerprint(s) already enrolled):

```
while(1)  {
    int result;

    result = HandlerReady ( 0 );
    TellKernel ( ex_Login() ? "AUTH-OK" : "AUTH-FAIL" );
}
```

The **ex_Login()** function below uses the global variable NB_OF_ATTEMPTS, which is set to the number of authentication attempts allowed a user to provide a live matching fingerprint before an authentication failure is returned to the kernel.

```
for(nb of attempts = 1; nb of attempts <= NB OF ATTEMPTS;
nb of attempts++)
   {
...

    if ( fingerprint authentication (fd) )
    {
      TellUI ( "FP:shw:User Authenticated" );
...
      plugin release device(fd);
      return -1;
    }
   }  // for nb_of_attempts
...
return(0);
```

The **fingerprint_authentication** function called by **ex_Login()** determines where to carry out the work (i.e., either off or on the PDA), as shown below.

```
int fingerprint authentication (int fd)
{
```

```
#ifdef FIU300 ONLY
// scanning and processing of the fingerprint done on the device
i = compare on fiu300(fd, 5);
#else
// scanning of the image, then transferred to the pda, which process it
i = compare on pda(fd);
#endif // FIU300_ONLY

if (i < AUTH THRESHOLD)
{
  printf("fingerprint authentication: failed\n");
  return 0;
}
printf("fingerprint authentication: success\n");
return 1;
}
```

For the lightweight authentication variant, the **compare_on_fiu300** function is called. The work is eventually performed on the fingerprint device, using **findIDInListwithScan()**, which is part of the system development kit for the fingerprint unit. The function builds the appropriate command according to the parameters given, sends it to the unit, gets acknowledgment from the unit, and returns the score.[2] The unit itself scans a fingerprint image, creates gray scale and monochrome data, and compares the monochrome data with template data at specified index numbers to obtain a matching score.

```
/************************************************************************
*/
  int compare on fiu300(int fd, int nb registers, ...)
  {
...

    if ( (i = findIDInListwithScan(fd, &score, 10, 8, 0X00, 0X17 0X00,
                                  0X02, 0X00, 0X00, 0X00, 0X01)) )
    {
       fprintf(stderr, "compare : findIDInListwithScan failed (%d)\n", i);
       return(-1);
    };

    return score;
  }
```

---

[2] An overview of the functions supported by the system development kit for the FIU-300 is available at
http://bssc.sel.sony.com/Professional/puppy/files/PUPSDK1.pdf

Fingerprint Authentication – Heavyweight

The heavyweight fingerprint authentication mechanism relies on the computational capabilities PDA to perform the entire authentication process. The PDA relies on the fingerprint sensor only to capture fingerprint images on demand, performing feature extraction, template storage, matching, and verdict resolution using its own computational resources. Figure 4 illustrates the functional organization of process, split between the fingerprint unit and the processor unit of the PDA. The term heavyweight is used to describe this approach, since most of the work is done directly on the PDA rather than the fingerprint unit, reducing the hardware needed for the fingerprint unit to a sensor.

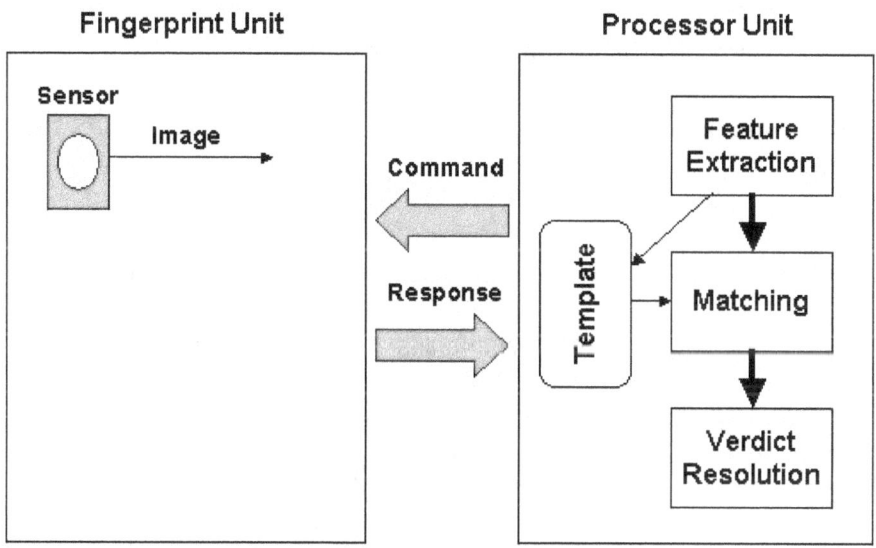

Figure 4: Heavyweight Functional Organization

The heavyweight solution can significantly overwhelm the processing load on the PDA processor unit. Two experiments performed with open source code, cross-compiled to the ARM[3] processor used on the iPAQ, illustrate this point. Executing an early version of a fingerprint verification system from SourceForge took approximately one hour to feature extract and match a fingerprint.[4] While an order of magnitude better, the NIST fingerprint image software took several minutes to perform the same task.[5]

The heavyweight application uses a commercial product for Original Equipment Manufacturers called FingerCell, which is specialized for handheld devices.[6] The FingerCell Embedded Development Kit (EDK) provides a software library, documentation, and a sample fingerprint database for developing an embedded fingerprint identification system. The FingerCell library functions include feature extraction, feature generalization, matching, and algorithm parameter setting. The library is designed to run on ARM-based platforms and for compatibility with the Arm-Linux GCC C compiler. An ARM based processor with at least 150 MHz Central Processing Unit (CPU) clock rate results in fingerprint enrollment in less than one second.

---

[3] More information on ARM processors can be found at http://www.arm.com/
[4] More information can be found at http://fvs.sourceforge.net/
[5] More information can be found at http://fingerprint.nist.gov/NFIS/index.html
[6] More information on the FingerCell library can be found at http://www.neurotechnologija.com/fc_edk.html

Overview

The fingerprint authentication process illustrated earlier in Figure 3 is also applicable for the heavyweight solution. The heavyweight variant of fingerprint handler, as with the lightweight variant, is a non-polling handler that runs in user space. It also manages the mechanism by communicating with the kernel, the fingerprint reader, and the Opie plug-in containing its user interface components in a similar fashion, guiding the fingerprint scanning process and controlling all the necessary steps during the authentication process and verdict determination. The main difference, however, is that once an image is captured from the fingerprint sensor, the handler relies on the algorithms of the FingerCell library to extract features and create templates and to perform matching operations.

Two different sensors were used in the implementation:

- The Fingerprint ID Unit (FIU) 300, discussed earlier in the lightweight fingerprint authentication section
- The Atmel FingerChip AT77C101B

The automatic template extraction, storage, and matching features of the FIU-300 were not used in the heavyweight variant. Instead, only its ability to capture and return an image was relied on. However, communications with the unit were encrypted using the FIU-300 supported features.

The FingerChip single-chip sensor is integrated into the body of iPAQ 5400's and 5500's and interfaces to the USB host controller.[7] It uses physical temperature effects for fingerprint sensing. The sensor comprises an array of 8 rows by 280 columns, giving 2240 temperature-sensitive pixels. The fingerprint image is captured by sweeping the user's finger across the linear sensing area. Sweeping captures successive images (slices) from which the fingerprint is reconstructed. Reconstruction produces a large, high-quality, 500 dots per inch image of the fingerprint. Unlike the FIU-300, no function exists to support encrypted communications with the processor.

Protection

For user authentication the fundamental threat is an attacker impersonating a user and gaining control of the device and its contents. Fingerprint units should be embedded into devices that are designed to resist physical tampering and avoid exposing the communications channel between the device and the unit. Presuming those safeguards are effective, the following vulnerabilities are the main candidates for exploitation:

- The authentication mechanism can be bypassed
- Weak authentication algorithms and methods are used
- The implementation of a correct and effective authentication mechanism design is flawed
- The confidentiality and integrity of stored authentication information is not preserved

As with the lightweight variant, the heavyweight fingerprint authentication mechanism relies on MAF, which in turns relies on the security of the underlying operating system implementation. The heavy fingerprint handler and its UI components are protected using the policy enforcement

---

[7] Information on the driver used for the Familiar distribution on the iPAQ can be found at
http://heim.ifi.uio no/~jorgenam/h5400/

mechanisms and policy rules as that of the lightweight variant. When communicating with the FIU-300, all messages are encrypted.

Handler Implementation

The heavyweight fingerprint handler operates as a non-polling handler, allowing a prescribed number of fingerprint scans before giving up. The main execution loop of the handler is the same as that of the lightweight variant discussed earlier. The **fingerprint_authentication** function called by **ex_Login()** from the main loop is also identical. However, instead of the work being carried out on the device, fingerprint authentication is carried out on the PDA through the use of the **compare_on_pda** function listed below.

```
int compare on pda(int fd) {
 ...
// get the collection of enrolled fingerprint files to compare the
fingerprint to
  n = scandir(WORKING DIRECTORY, &enrolledList, filter, alphasort);
...
  // Put the finger on the sensor
...
  if ( (i = scan(fd)) )
   ...

  TellUI( "FP:shw:Scan complete\nPlease remove your finger");

  // copy the grayscale from the device to the pda
  if ( (i = getGrayscale(fd)) )
   ...

  // show the image to the user
  if( !(buffer1=strcatalloc(0, 3, "FP:shwimg:", WORKING_DIRECTORY,
"grayscale.pgm")) )
   ...
  TellUI ( buffer1 );

  // Processing the file
  // first, the image of the fingerprint just scanned is of grayscale type
  type1=0;
...
  if (!(buffer1=strcatalloc(0, 2, WORKING_DIRECTORY, "grayscale.pgm")) )
   ...

  while(n--)
  {
// determine the type of the file which will be processed
    if ( !(i=fnmatch("*grayscale*", enrolledList[n]->d_name,
FNM NOESCAPE)))
    {
      //grayscale
      type2=0;
    } else {
      //template
      type2=1;
    }

    if ( score < AUTH_THRESHOLD)
```

13

```
   {
// compare with the next file in the collection we found, only if
necessary
     if (!(buffer2=strcatalloc(0, 2, WORKING_DIRECTORY, enrolledList[n]-
>d name)) )
       {
         //error handling
       } else {
         if ( (score = compare fingerprints(buffer1, type1, buffer2, type2)
) < 0)
         ... //error handling
       }
     }
     free(enrolledList[n]);
   }

   /* delete the raw image */
   printf("Deleting the grayscale image got from the device for comparison
(%s)\n", buffer1);
   i=remove(buffer1);
   ...

   return score;
}
```

The **compare_on_pda** function collects a live fingerprint and then compares it to the enrolled set of fingerprints using the FingerCell **compare_fingerprints** function. Similarly, during enrollment, the heavyweight solution uses a FingerCell function to generalize the fingerprint images scanned to remove noise and extract the common characteristics to create a template.

Management interface

The fingerprint management interface is the same for both the lightweight and heavyweight solutions. It can be launched at any time to enroll or update the set of enrolled fingerprints via an icon located on the tab "Settings" of the Opie desktop. If fingerprints have been enrolled, the user must authenticate to the mechanism with a live fingerprint, before any changes are allowed. The snapshots of the user interface are shown in Figure 5 below, showing the attempt to enroll a new fingerprint (at left) and delete an existing one (at right).

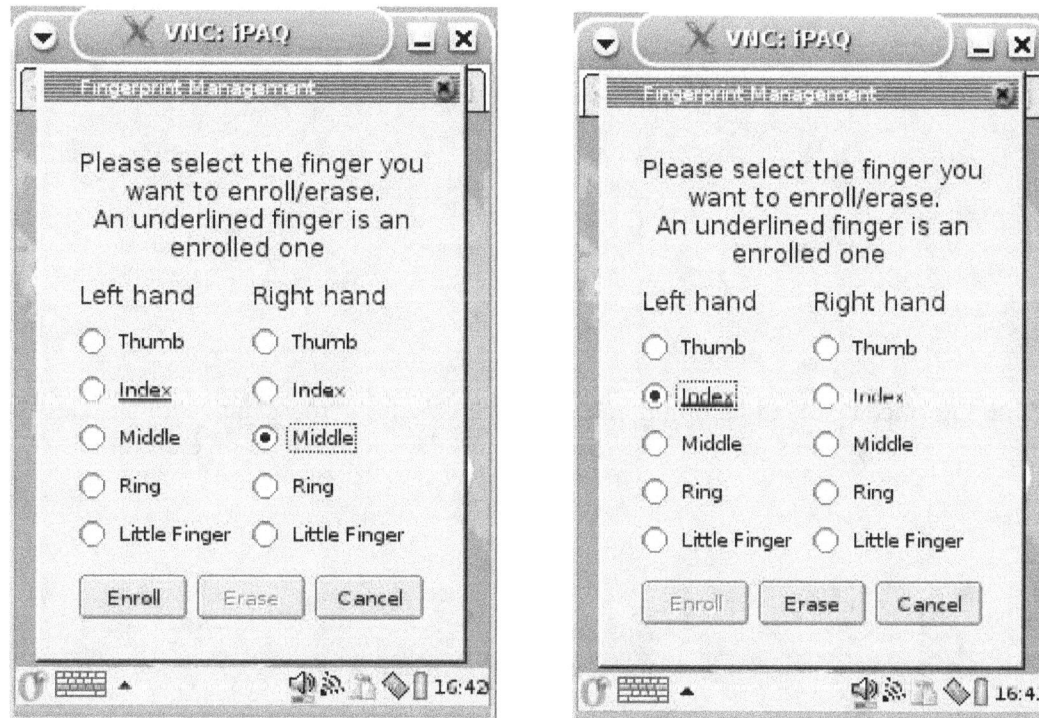

Figure 5: Fingerprint Management Screens

The management interface is very simple and intuitive to use. The user can select any finger of either hand. For the selected finger, an action is proposed corresponding to the enrollment state of this finger. The user can erase the information using the Erase button or add a fingerprint using the Enroll button. :

- If the finger is enrolled, its name is underlined. If the user chooses to erase a fingerprint, all information concerning the fingerprint is cleared, after a confirmation. If the user chooses to update a fingerprint, the existing information must be cleared before the enrollment process can proceed (i.e., the finger must not be enrolled).
- If the finger is not enrolled, no underline appears and the user can enroll the finger. Three consecutive fingerprint scans are taken, which are compared against one another using a round robin protocol. The one with the best average result is selected as the representative template for the user's finger.

After an action is completed, the interface is updated and the user can continue with any additional fingerprint management functions.

The fingerprint management interface is contained in the Opie plug-in, with the other MAF interfaces. The graphical part is written in C++ and completely separated from the processing part, the handler itself, written in C.

The dialog between the handler and the interface is done using the dialog protocol, as shown in the following code with the erase_fingerprint() method:

```
// This method is called when the user presses the Erase button, after
choosing a finger
int FPManagementUI :: erase fingerprint() {

// handFinger is the finger currently selected

 char Seq[250];
 snprintf(Seq, 250, "FPM:eras:%i", handFinger);
 if ( -1 == msgMux->SendMsg(Seq, (struct sockaddr*)rAddr, rAddrLen) ){
       perror ( "Error Sending the sequence" );
 }


}
```

Every time the interface needs to communicate with the handler, a message is constructed and sent. The handler analyzes this message, as illustrated in the following code example:

```
// Test if the message is an erase command

if ( strncmp ( msgFromUI, "FPM:eras", 8) == 0 ) {
       printf("erase_fingerprint command received \n");

       char str [10];
       int tp_fing = 0;

       // We scan the message to get the finger number
       if( sscanf ( msgFromUI, "%9s%d",str,&tp_fing)){
             printf("\nerase fingerprint - sscanf done, finger = %i \n" ,
tp fing);
             // we launch the erase process
             if (!( erase fingerprint(tp fing) )) {
                   sprintf(msgToUI, "FPM:quit:%i", 4);
                   TellUI(msgToUI);
                   release device(fd);
                   return(-1);
             }
```

References

[Beo02]   Nicky Boertien, Eric Middelkoop, Authentication in Mobile Applications, CMG,
           Telematica Instituut, The Netherlands, January 2002, <URL:
           https://doc.telin.nl/dscgi/ds.py/Get/File-23314/VH_authenticatie.pdf>.

[Jan03]   Wayne Jansen Vlad Korolev, Serban Gavrila, Thomas Heute, Clément Séveillac, A
           Framework for Multi-Mode Authentication: Overview and Implementation Guide,
           NISTIR 7046, August 2003, <URL: http://csrc.nist.gov/publications/nistir/nistir-
           7046.pdf>.

[Pol97]   Despina Polemi, Biometric Techniques: Review and Evaluation of Biometric
           Techniques for Identification and Authentication, Institute of Communication and
           Computer Systems, National Technical University of Athens, April 1997, <URL:
           ftp://ftp.cordis.lu/pub/infosec/docs/biomet.doc>.

[Ulu04]   Umut Uludag, Anil K. Jain, Fingerprint Minutiae Attack System, The Biometric
           Consortium Conference, Arlington, VAm September 2004, <URL:
           http://www.wvu.edu/~bknc/2004%20Abstracts/Fingerprint%20Minutiae%20Attack%
           20System.pdf>.

Appendix A – Software Organization

The fingerprint authentication software is organized into the following components:

- FPManagementUI.cpp - This is the management interface class, where enrolling and deleting fingerprints are implemented.

- main.c - This is the regular main code of the handler that is launched by the kernel to carry out authentication. The **ex_Login** and **fingerprint_authentication** functions discussed in the body of the report are part of this component.

- fingerprint.c – This Contains the main methods of the fingerprint handler, such as reading/saving configuration, initiating comparisons, etc. The **compare_on_pda** and **compare_on_fiu300** functions discussed in the body of the report are part of this component.

- authentication.c - The methods here carry out the authentication. They are called by the methods in fingerprint.c. If the authentication library or its methods are changed this part should be re-implemented.

- encryption.c - The FIU-300 uses encryption for communication; the methods used to perform cryptography using the openssl library are embodied here.

- serial.c –The generic methods that fingerprint.c is calling and also specific ones for the specific device being used are located here. This procedure should be re-implemented if another RS-232 serial scanner is used. The heavyweight variant relies too much on the specific communication protocol of the sensor to abstract out generic functions. For the lightweight variant, the generic methods that would need to be re-implemented for another scanner are:

  - getGrayscale()
  - release_device()
  - initialise_device()